Managing and Marketing Sheep

Tools and Techniques for Every Shepherd

By
Darla Noble

Mendon Cottage Books

JD-Biz Publishing

Table of Contents

Introduction

Conclusion

Author Bio

Introduction

The keys to raising sheep for profit lie in **management** and **marketing**. Managing your flock to achieve optimal health, growth and productivity in the most cost-productive manner is absolutely essential if you wish to market your animals for the best possible price. Rather simplistic sounding, isn't it? It can be—if you follow a few basic guidelines and remember:

You can't starve a profit out of your animals.

No one has as much to gain or lose as you do from managing and marketing your flock, so no one should work as hard as you do.

Both of these statements should be at the core of your business plan (yes, farming on even a small scale should be viewed as a business) and the purpose of this book is to help you a) get to that point or b) give you an 'atta boy and a few ideas to take you to the next level of production and profit

Chapter 1: Getting Off to the Right Start

The first step in proper flock management is to have healthy, hardy sheep to manage.

If you are just starting out, this means purchasing quality stock from a reputable producer. It also means NOT buying your stock from a sale barn or from someone who promises the moon and tells you the sheep will raise and take care of themselves.

It's worth paying a little more for quality animals—even if it means you have fewer animals to begin with—than to buy more sheep for the same amount of money that are of inferior quality.

Quality ewes and lambs

If you already have a flock of sheep, you need to take a good look at what you have and possibly cull those that don't meet the standards you wish to maintain. When deciding who can stay vs. who should go, ask yourself:

*Do they twin or triplet after the first lambing

*Do they produce enough milk to raise healthy, weighty lambs

*Are they efficient with their feed

*Breed easily

*Are they resistant to parasites—rarely, if ever, needing to be treated for parasites

*Do they sufficiently maintain their weight throughout lactation

*Do they display strong confirmation and breed standards

These girls can stay. But if the following characteristics plague any of the ewes in your flock, they need to go…ASAP:

✓Produce only single lambs

✓Display poor mothering instincts or reject their lambs

✓Abort or give birth to dead lambs

✓Contract pink eye or foot rot

✓Prolapse*

✓Low resistance to parasites

✓Have trouble maintaining their weight sufficiently during lactation

✓Take more than 2 cycles to breed

✓Rogue in nature—refuses to be herded without a good amount of difficulty

✓Consistently gives birth to low birth weight lambs who don't grow well

✓Display substandard breed characteristics and confirmation

These girls need to go. What's more, unless you breed strong genetics into their ewe lambs, you probably need to resist the temptation to retain them into your flock, too. I know…it sounds harsh, and there are exceptions, as noted below, but remember…this is business.

Three year-old ewe and her triplets

One of co-author, Darla Noble's strongest production ewes consistently gave birth to strong, healthy twins. One year, however, she gave birth to a single ewe lamb...a large ewe lamb with some physical deformities including blindness in one eye. While the lamb was obviously not going to be kept in the flock, the Nobles weren't willing to count the mom out because of one freak accident. The decision was a sound one, as she went on to produce quality lambs for several years to come.

*If a young ewe experiences a prolapse, you need to examine why it happened. Are your feeders too high; making it necessary for the ewes to climb to eat, which allows gravity to put excess strain on their vaginal wall? If so, correct the situation. *This* is an example of proper management. Did her mother prolapse?

This indicates a genetic disposition to the condition you don't want in your flock. How old is the ewe? Younger ewes who prolapse will likely do it again so should be culled. Older ewes who prolapse for the first time may have been stressed or injured. If the prolapse is easily repaired and no other problems occur, you should weigh the situation; taking into consideration her past performance before deciding what to do.

Now for the other half of your flock…the ram. Yes, I know my math, but the ram is half the flock in the fact that he can easily breed up to fifty ewes. So you see, half the lamb's genetics comes from that one ram.

Mature ram

When selecting a ram you need to take the following into consideration:

✓ Confirmation which exhibits strong breed standards.

✓ Birth type-your ram should be a twin.

✓ Black-hoofed (while not a 'must', it will cut down on the need for hoof trimming).

✓ You should never buy a ram younger than 12 to 18 months old. Purchasing a ram lamb is a huge mistake, as you have no idea whether he'll 'fall apart' when weaned—only to be unable to recover satisfactorily.

✓Buy off the farm. If possible, buy a 2 to 3 year-old and ask to see lambs he's produced.

Ram lamb…there is no way to know how he will grow out at this stage in his life.

Once you have a quality flock of sheep to work with and who will work for you, it's time to get down to the business of managing them in order to sustain the quality necessary to take the fullest possible advantage of the market.

Chapter 2: Feeding Ewes and Rams

The quality of your flock begins with good genetics, but it will only continue if you have a feed program that works to maintain and build upon what you have.

Let me begin this by saying that there are some producers out there that brag about the fact that their flock is a low-maintenance one. Unfortunately their definition of low-maintenance is no-maintenance…and it shows.

Feeding sheep is not a one-size-fits-all kind of thing. What you feed, how much you feed and even when you feed depends upon where a sheep is in its 'life cycle'.

EWES

What to feed: Ewes need a steady and ready-supply of grass or hay accessible to them 24/7. The roughage is necessary for their digestive systems as well as the nutritional value of many of the grasses. It is important, however, to remember that all grasses are not alike. And since they live on pasture, it is naturally comprises the vast majority of their diet.

Ewes do best on cool-season grasses; fescue, ryegrass, orchard grass, Kentucky bluegrass and timothy to name a few, with fescue being the most common in the US. The cool-season grasses contain varying amounts of selenium, protein and other essential vitamins and minerals. Fescue is usually naturally reserved for winter foraging by the sheep themselves. Fescue contains a fungal endophyte that is bitter and unpalatable until a hard frost kills the grass; making it a standing forage the sheep can enjoy.
Warm-season grasses include Bermuda, switchgrass, crabgrass and sudan grass. The sheep can survive on them just fine, but traditionally do better on cool-season grasses or pastures that contain both cool and warm-season grasses.
And finally…legumes. Legumes include alfalfa, clover, vetches, trefoil grasses and lespedezas. Legumes are filled with proteins which

are fine…within their proper limits. Too much protein at certain times is toxic to sheep and can actually damage or inhibit their fertility. Cool…warm…legumes…the best pastures is those that contain a mix of all three. The best pastures are also those that are not eaten into the ground, but are allowed to rest and replenish themselves as a result of rotational grazing (we will 'talk' more about that later).

But just like man cannot live on bread alone, sheep cannot live on grass alone (in most cases, anyway). At certain stages of a ewe's life cycle, they need additional energy (carbs) which usually comes in the form of corn. Other sources of energy for ewes can come from oats, cottonseed meal and barley. Corn, however, is usually the most cost-effective so is the most commonly used source of energy for ewes. Protein-filled feeds give ewes the ability to fight off parasites, and provides the necessary micro-organisms for proper rumen function. Sheep don't store protein in their bodies—it's burned as energy (not as efficiently as carbs) and nitrogen. Protein comes in the form of alfalfa pellets, distillers grains, soybean meal and brewer's grains, to name a few.

Mineral mix is the other aspect of a ewe's diet that should be available to them 24/7. The mineral mix should be **specific to sheep**—meaning:

*It will be loose rather than in block form. Mineral blocks are hard on a ewe's teeth; causing breakage, which in turn, decreases their lifespan and hear of productivity.

*It will contain no copper. Copper is toxic to sheep. There are several products on the market that claim to be cross-species compatible. Don't believe them. Mineral mixes containing copper should not be fed.

When to feed what: By and large, ewes will thrive on quality pasture/forage, mineral mix and water the majority of the year. This means that unless you are experiencing drastic and severe weather conditions, these three 'food groups' are all that is necessary.

So what constitutes weather extremes?

Extreme cold (below-zero) and cold combined with heavy snow or ice. In weather like this your ewes need extra carbs (energy) to keep their bodies generating the heat they need to stay warm.

Severe heat and drought. The severe heat and drought wears sheep down and leaves them highly susceptible to parasite infestations. Extra protein is called for at times like in order to give ewes the ability to fight off parasites and keep their digestive systems in top working order.

Other than extreme weather conditions the times in which you need to feed your ewes something besides forage are:

One: Prior to breeding. If your ewes are a bit on the thin side (for whatever reason) you may want to begin feeding them about a pound of shell corn apiece for 4 to 6 weeks prior to breeding. This is called 'flushing'. The added calories is said to increase the number of eggs release for fertilization, which obviously results in more multiple births. However…sheep who are in good condition should not be flushed, as the added weight gain on a sheep who doesn't need the extra pounds can actually hurt their chances of breeding at all.

Ewes on late summer pasture of cool season grasses

Two: The last few weeks (2 to 4) of gestation through a couple of weeks prior to weaning their lambs from them. A few weeks prior to lambing until a week or so before weaning their lambs, ewes need approx. 2 pounds of feed apiece per day. The feed can be shell corn, distillers grains pellets or a mix of the two.

NOTE: If you are fortunate enough to have a pasture that contains a good mix of grasses and legumes and that pasture is lush enough to amply support the entire flock during this part of their life cycle, you may be able to forgo feeding supplemental grains.

Ewes on summer pasture still need supplemental feed if not high quality grass.

Winter lambing ewes eating hay

RAM:

The Ram Diet. No, it's not the latest craze in weight loss mania, but the term we'll use to talk about what and how to feed your ram to keep him in top condition and to get the most from him when it comes time to breed your ewes.

Feeding the ram is actually quite simple. You need to keep your ram conditioned to a body score of 3, which will can usually be accomplished with free-grange grazing of quality grass or 4-5 pounds of hay per day. Some shepherds prefer to feed their ram a half a pound of corn or pelletized feeds per day during breeding season to provide extra energy. But since the ram spends the majority of his time alone, it is usually not difficult to keep a ram in top condition on pasture alone; reserving grain feeding for the harsh winter months if necessary.

Like your ewes, the ram also needs 24/7 access to sheep mineral mix and water.

That's about it for the rams. I know…I almost feel negligent or biased in giving pages to the feeding of ewes and less than a page for rams, but it is what it is and there's no use beating a dead horse, now, is there?

I do want to say this, though, the fact that ram nutrition is not as varied as that of ewes (and lambs, as you are about to learn), doesn't mean it isn't important. Keeping your ram well-fed and healthy ultimately plays a major role in the health and well-being of your entire flock.

Chapter 3: Feeding Lambs Correctly

When it comes to feeding lambs—ewe lambs and ram lambs—there are certain elements that cannot be ignored if you intend to a) grow replacements for your aging ewes and/or add to your flock and b) sell quality breeding stock to other breeders and market lambs for processing.

Spring/summer ewe lambs being fed after weaning for optimal growth before coming of age to breed

The first thing you need to take into consideration when planning a feeding program for your lambs is the time of year in which they are born. And the time of year in which you lamb will be dependent on the markets you are raising your lambs for. FYI: More on that later.

Lambs born in the spring or summer will be able to be raised on their mother's, quality grass and mineral with supplemental grain being fed for finishing them out for slaughter (ram lambs).

Producers who lamb in the fall and winter, however, will need to begin feeding small amounts of grain a couple of weeks after their born and <u>slowly</u> increase their feed rations for optimal growth. This is, of course, in addition to their mother's milk and hay.

Let's take a closer look at each 'option' and how to ensure your lambs are growing to their full potential with each one.

SPRING/SUMMER LAMBS:

Lambs born March-June will spend their first two months nursing from their mother and grazing on the first grass of spring. As long as the grass is quality grass and there is plenty of it, you can easily raise healthy, hardy lambs. You will, however, need to worm the lambs at weaning and then again thirty days later. The stress of weaning can explode the parasite population in a lamb's system; stunting their growth and ruining their potential.

Ewe lambs can easily acclimate into the same feeding/foraging system of adult ewes and be ready for breeding when they come of age.

Ram lambs, however, must be ready for market at the most opportune time, i.e., when they will garner the biggest profit. This will require you to increase their protein intake by feeding grain—usually a pelletized feed for lambs or protein-rich grains like soybean meal or distillers grains with a 15% to 17% protein content.

Feeding out your ram lambs should be done gradually and consistently in order to maintain steady growth. Feeding too much too quickly will upset their digestive systems and set them back; leaving you with lambs to sell that aren't the quality they could be.

SPRING/SUMMER RAM LAMBS:

Upon weaning, begin offering small amounts of grain to the lambs; a handful of grain per lamb once a day. They will probably not eat the grain the first few days, but once they know what it is, this won't be a problem. When this happens, increase the amount of grain to 2% of the average weight of the lambs. EXAMPLE: Weaned lambs will average 20 pounds or so in weight. This means each lamb would need ¼ pound per lamb once a day.

Every 3-4 days increase the amount of grain given based upon the fact that lambs on grain will gain an average of a half a pound a day. EXAMPLE: Increase the feed from ¼ lb. per lamb to just under ½ lb. of grain per lamb once a day. Continue this until you are feeding 2 lb. per animal per day. Maintain this ration until your lambs average the desired market weight.

FALL/WINTER LAMBS:

Fall and winter lambs are those lambs born between August-February. These lambs-both ewes and rams-will need supplemental grain feeding if you a) intend to have the lambs ready for the meat market when they need to be or b) you want your ewe lambs to get off to a great start in order to be ready to breed when they are a year old.

When feeding fall/winter lambs follow the same feeding regimen as you do with spring/summer ram lambs, but with the following changes and additions:

*Begin the feeding lambs within the first couple of weeks of life

*Feed lambs separately from the ewes to ensure the lambs are getting what they need

*Use hay feeders that will allow lambs to have access to all the roughage/forage they want

In feeding out your lambs, you need to understand that the goal is consistent, steady growth. Trying to push too much feed too fast will upset their digestive systems; causing scours and stunted growth. Trying to get by on less feed in order to save money will result in thin lambs that don't butcher out well; meaning the amount of meat is not what it should or could be.
Or in the case of ewe lambs, they will not have the body conditioning necessary to produce strong lambs and enough milk to raise them adequately.

Fall/winter lambs learning to eat grain and hay

Chapter 4: Let's Get Down to Business

At this point your flock consists of quality animals which, in turn, produce quality lambs for replacement ewes and market lambs. But raising them isn't enough. If you intend to make your farm a profitable business venture—not matter how big or small you farm is—you need to know how to market your lambs when and where it will be most advantageous to you.

EWE LAMBS:

Ewe lambs for a starter flock

Ewe lambs are marketable just about any time of the year. In other words, when you have a group of ewes to sell—lambs or adults—you should have no problem selling them if people know you have them. These people may already be raising sheep and looking to add to their flocks or they may be people who are just starting out and in need of a quality starter flock of ewes.

You will find it profitable, however, to have a small group (or two …or three) of one to three year-old ewes for sale when you attend any sort of agricultural event at which you will have an opportunity to network with other people interested in farming.

People spend years getting an education in the how's and why's of marketing. Now while I am in no way discounting the time and effort these people put into their education, I *am* telling you that marketing is not rocket science. If you have a product to sell (in your case, the product is sheep) all you need to do is equip yourself to be:

✓Knowledgeable about your animals—not only their breed standards, but the facts about a particular animal.

✓Honest. 'Nuff said.

✓Approachable—greet people first. Smile, say hi and make contact. Talk to other people about their sheep; show interest in what others are doing.

✓Educational. If you have the opportunity to do so, have a simple, yet informational display of your farm. This should include pictures of your flock, information on the breed, mints or hard candies and most of all…business cards with your contact information on them.

✓Interactive. If you are displaying your sheep in a pen at a fair or farm expo, keep the sheep clean and keep the pen clear of chairs, etc. so people can get up-close and personal with them. BUT…do not allow people to feed or taunt the sheep. Talk to passers-by. Find out if they are interested in raising sheep and if so, answer any questions they by giving honest, brief, yet thorough answers.

Farm Expos make for great networking opportunities

✓ Available. You might have the best sheep this side of the Mississippi, but if you don't let people know you have them for sale, your sheep are worth absolutely nothing. You've got to spread the word.

When you are able to present yourself and your sheep in such a manner, if you have sheep to sell, you will sell sheep.

RAM LAMBS for SLAUGHTER:

Ram lambs require a different type of marketing. You are selling poundage rather than breeding stock. You are selling for consumption instead of production. You are also working with windows of opportunity—strict time frames for getting the highest price for your lambs. In other words, if there is one word that could be used to define the art of marketing slaughter lambs that have been fed and managed properly, that word would be 'timing'.

There are three prime markets a producer should focus on to have slaughter lambs ready to go to market. They are:

EASTER—Christian and Greek Orthodox (sometimes several weeks apart). Lamb is the meat of choice for many celebrating Easter. For this, consumers want a small lamb for roasting/baking. The best prices will be paid for live lambs weighing between 45-60 pounds. Because Easter is celebrated in early to mid-spring, this means you will need to lamb between December and early February in order to have lambs at the desired weight at that time.

I'm going to digress just a moment here to have you think back to what was discussed in the feeding of fall/winter lambs. Feeding begins just days after birth rather than at weaning like it is for spring/summer lambs. Now while this does mean spending money on feed earlier, you are only taking them to 50 pounds or so (smaller than other desired markets). So when you sit down and do the math, you are making a profit.

RAMADAN—an Islamic holiday celebrated at the end of the month of Ramadan. Because Islamic holidays are lunar; meaning they count their days by the phases of the moon, this as well as other Islamic holidays travel across our traditional calendar. For example: in 2014, Ramadan is celebrated from the end of June to the end of July. Some years, however, it is as late as November.

The desired live weight of lambs being sold for slaughter for Ramadan celebrations usually averages between 65-75 pounds. In order to have lambs ready for this market, you will need to plan ahead knowing a year in advance when the holiday will be celebrated and breed your sheep to lamb accordingly.

FESTIVAL of the SACRIFIC—another Islamic holiday which 'travels', the premium price is paid for lambs weighing 75 to 80 or 90 pounds. Again, you will have to plan ahead to meet this demand.

While this is by no means the only time you can sell your lambs, these are the times in which you will receive the best price for a quality animal.

If yours is a small farm, you assume your only option for selling your lambs is the sale barn. In some instances, this may be true. But before you write this off as your only option, take a few minutes to consider a few other possibilities...

Farm sales: Advertise butcher lambs for sale on local websites, bulletin boards, newspapers, circulars you farm website, farmer's markets and other such venues. Sell by the pound (going rate) live-weight and deliver the animal to the local processor for them (if necessary) along with an order sheet to have the lamb processed to their specifications (stew meat, ground lamb, chops, etc.)

Co-ops: Band together with other small producers to sell to packing houses, restaurants, or order buyers.

Join with other producers to sell direct to the processing plant.

Why bother with taking a more direct sales approach to marketing your lambs for slaughter? The word 'direct' is your answer. Any time you can cut someone out of the channel a lamb takes to get from the farm to the table, you are putting more money in your pocket. Skipping the sale barn saves you commissions and fees that take a considerable piece of your profits. Think about it—why should you give someone else part of your profit for selling sheep you are capable of selling yourself?

RAMS for BREEDING:

Selling rams for the purpose of breeding stock should be one of the most labor-intensive aspects of raising sheep. Rams sold for breeding should be raised and fed to be top-notch examples of their breed and body conditioning. Rams sold for breeding should be in perfect health and come from the best genetics in your flock.

The problem comes when sheep producers have the erroneous idea that just because a ram is a ram, he is a breeding ram. NOT TRUE! I would feel confident in saying that somewhere between 0 and 3% of ram lambs born will mature to be suitable breeding rams.

If, however, you have rams that fit the bill for breeding, make sure you can provide proof of genetics, birth type, etc. to potential buyers. And whatever you do, PLEASE don't try to sell a ram lamb as a breeding ram. When selling breeding rams, you should not do so until they reach one year of age. To sell a ram lamb as a breeding ram is like selling a car without a transmission…without telling the new owner it's missing.

As for how to market breeding rams, your farm website, breed association newsletters, meetings and ag expos are your best options.

You'll notice I've not said anything about pricing sheep and lambs. The reason for this is the fact that prices fluctuate from year to year, region to region, breed to breed and farm to farm. Pricing your animals is something you are going to have to come to on your own by watching the market (for market animals) and talking to other breeders.

Chapter 5: Techniques and Tools for Marketing

In the previous chapter we touched briefly on the value of networking, being available and being knowledgeable about the product you are marketing/selling. Now let's take a little more time to take a closer look at these tools and techniques—things you can do to sell your sheep as efficiently and profitably as possible.

Business cards: Never underestimate the power of a little business card. Hand them out at meetings, farmer's markets, the feed store, anytime you get the chance to talk about what you do.

Website: No matter how big or small your farm, you need to have a website. The world-wide web isn't called world-wide for nothing. Before retiring, Generation 5 Farm, which was located in mid-Missouri, sold sheep all across the United Sates...because they made themselves available on the internet.

Your website should be simple, yet paint a complete picture. It should contain pictures of your sheep/farm, contact information and facts about sheep care and management. Websites can be created at no cost, but purchasing your name is not expensive. Either way...a website is a must!

Associations: Belonging to your breed association, your state's sheep producers association, various agricultural associations (local, regional, state and national) and even your county fair board are invaluable marketing tools. Not only will you be getting your name out there, you gain access to grants, incentive programs and other resources to enhance your farm.

Community involvement: Being involved in 4-H and FFA not only allows you to help educate future farmers, but who do you think these kids are going to call when they need a market lamb or a starter flock of ewes? Exactly. You can also make money marketing your sheep without them never leaving the farm by marketing the farm as an agri-tourism stop; a place kids and adults alike can come to learn about sheep. Offer to talk to different groups about what you do.

Advertising: Advertising these days is much more than an ad in a newspaper or on the radio or television. Besides, I can't think of anything more *ineffective* than advertising sheep for sale using any of these. Instead, you need to think beyond traditional. Think t-shirts, hats and jackets with your farm name on them. Wear them…sell them…get your name out there with them. Magnetic signs on your car or truck with your farm name, the breed of sheep you raise and your phone number/website are a great way to advertise. I've seen people writing down my contact info while sitting in traffic more times than I can count. And what's more, I've gotten calls and emails from people who say that is how they found out about our farm. Put a sign at the entrance of your driveway with your farm name on it and contact information.

Think outside the box: This is especially true when you are thinking about WHO to sell sheep to. Contact ethnic groups in your community, a near-by university or restaurant. Contact health food stores and groups who prefer buying local products and those that are farm-raised rather than commercially raised. What about the locally-owned upscale café? Selling to them for specialty dishes can be a win-win for everyone. NOTE: Just be sure you cover your bases with health regulations and codes.

Sheep for sale

Conclusion

Now tell me the truth…does any of this sound too difficult? I'll answer that one for you. No! Nothing you've read is difficult or unattainable. If you are committed and put some time and effort into planning and putting yourself out there, you can do raise, manage and market your sheep successfully and profitably.

Author Bio

Darla Noble is a native of mid-Missouri where she lives with her husband of thirty-three years, John. Darla's love of writing began in the fourth grade; after meeting up and coming children's author, Judy Blume,
who, by the way, autographed Darla's copy of "Are you there, God...it's me, Margaret".

Darla's love for writing and family makes her work sought after in the Christian market, parenting and family resources and ghostwriting for educators and inspirational speakers.

Health Learning Series

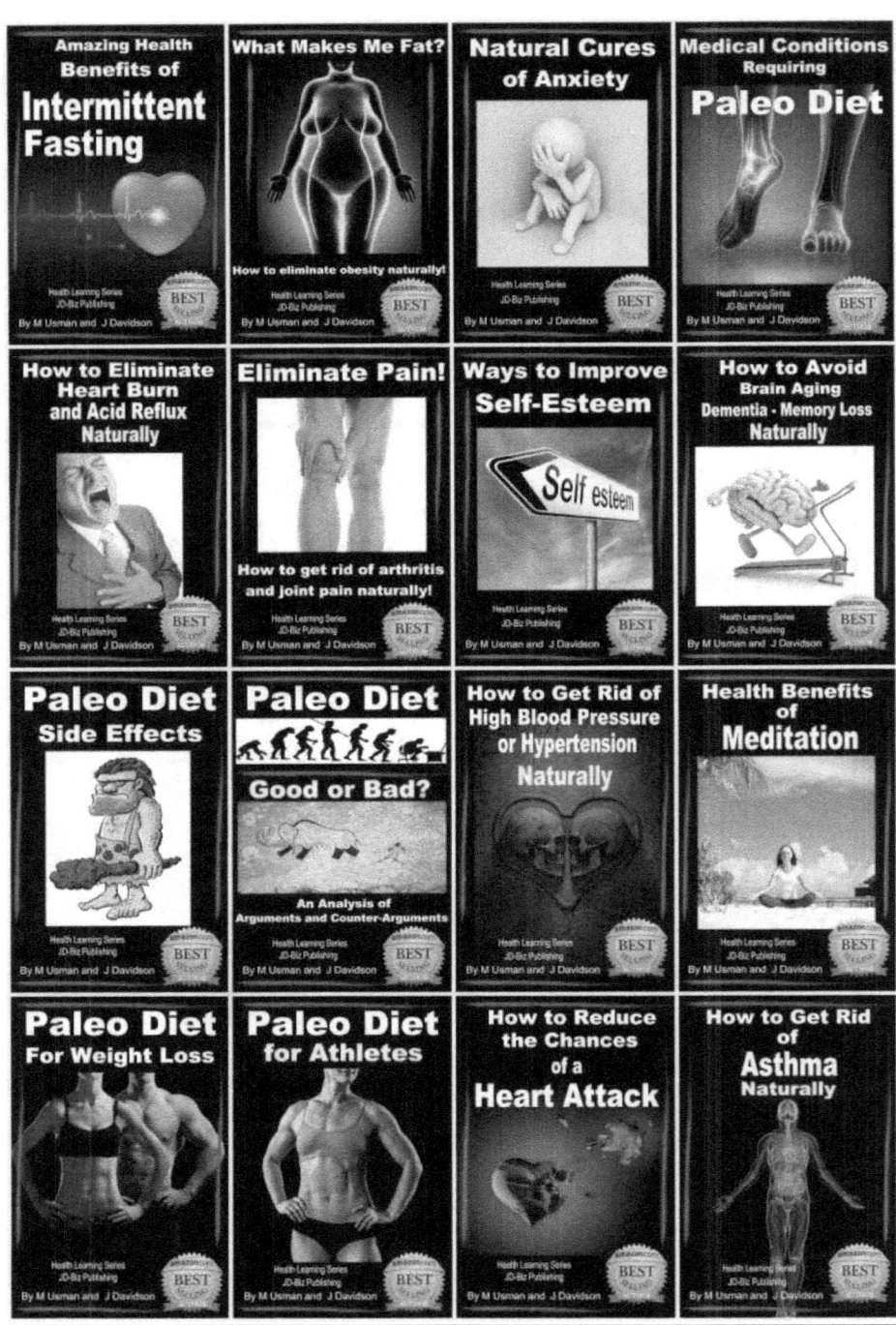

Learn To Draw Series

How to Build and Plan Books

This book is published by

JD-Biz Corp

P O Box 374

Mendon, Utah 84325

http://www.jd-biz.com/

www.ingramcontent.com/pod-product-compliance
Lightning Source LLC
Chambersburg PA
CBHW070726180526
45167CB00004B/1641